The information provided in this book is designed to provide helpful information on the Pick 3. This book is for entertainment purposes only. No part or parts of this book shall be copied or used without the sole consent of the author. Guides or work outs are provided for informational purposes only and do not constitute endorsement of any lottery websites or other sources. Readers should be aware that the author is not employed by any gaming companies now or in the past.

This chart was created for those learning the pick 3 circle the number that falls and play the number beside it on either side. Its used as a guide

060 061 062 063 064 065 066 067 068 069

240 241 242 243 244 245 246 247 248 249

150 151 152 153 154 155 156 157 158 159

080 081 082 083 084 085 086 087 088 089

350 351 352 353 354 355 356 357 358 359

170 171 172 173 174 175 176 177 178 179

260 261 262 263 264 265 266 267 268 269

190 191 192 193 194 195 196 197 198 199

280 281 282 283 284 285 286 287 288 289

370 371 372 373 374 375 376 377 378 379

460 461 462 463 464 465 466 467 468 469

390 391 392 393 394 395 396 397 398 399

480 481 482 483 484 485 486 487 488 489

570 571 572 573 574 575 576 577 578 579

low sum numbers

020 021 022 023 024 025 026 027 028 029

130 131 132 133 134 135 136 137 138 139

040 041 042 043 044 045 046 047 048 049

590 591 592 593 594 595 596 597 598 599

****

When one hits another will follow within 3 draws

479 047 125 146 167 235 245 248 249 257 278 279 467 478 012 015
018 019 023 024 027 035 038 039 044 045 048 049 057 077 078 079
123 128 129 135 138 139 158 159 224 227 228 229 234 237 238 239
247 289 345 348 349 357 358 359 378 379 446 448 449 458 459 489
578 579 677 778 779 789

****

Lets look at the number that fell last night 838 can we use the
doubler system yes it will stop when you get a 76/67 pair either
in a double or a single version.

This number sets off pairs that will fall and bounce off each
other all week . The pairs for the entire week and there may be a
draw it skips so play play play what you feel is coming....

The 9 pair set that is two numbers that equate to 9 =9,
09,81,72,63,54, outside the box 04x (we treat 0 as 5)

The 10 pair set = 10,55,91,00,64,82,73 and the most important set
of all 11 pair set b/c you got an 11 last night with 83 so

011, 55, 92, 11, 83,74,65 (do not stop this set just b/c they throw
you a sum 11 it will continue on)

****
these will hit within two draws of each other

346 , 351, 578, 537, 248, 864,489, 209, 237, 213, 151, 257, 704, 247,
260, 204, 466, 467,499,497
****
When one of these fall

847

658

027

145

Doesn't matter which way the number falls look for one of these pairs

41 ,51, 71

****
830/ when 830 shows a double of 00/11/22/33/44/55/66/77/88/99 will show soon look at the immediate double that fell the last time this number fell and you will go up one or down to get the next double so if the number was a 499 then you will look for a 55/00 double to show next time....research you will see what i mean

****
420/979 this does not always work especially if 420 has been a repeat from the month before normally yes if its awhile since that 420 came.

965/041

971/561

106/984/273 either one

499/214

332/866

****

Just a little reference to use when you think a double zero is coming have you seen this pattern in the numbers that fell ? Usually its a strong indication a double 00 is about to hit .

001

255

477

use when a 00 comes

****

350 and family .... reference above but be sure to note that 350 is also kin to 365 , 566, 366 one of these numbers will fall within 14 draws
****

****

010/ just off topic here anytime you see 00 you will see the 8's sooner than later so its like paul revere the 8's are coming the 8's are coming...80,81,82,83,84,85,86,88,87,88,89 I know its a lot of sums but you can reduce it by the last 8x so if the last 8x was 83 play the one above or below it .

****

Anytime you get one of these numbers and it doesn't matter which way it falls it will set off a chain reaction of pairs .

701

702

700

711

722

712

800

811

822

801

911

922

820

109

209

129

you will have a chain reaction of these pairs

49x 94x

27x 72x

38x 83x

16x 61x

05x 50x = 55x

I call them running mates anytime you see one fall another is sure to follow

350-805-358-459-409

540-349-348-803-389-

905,489

Anytime you see one of these numbers a pair 48x or 96x will show soon

120 112 180 196 336

****
3,6,9,33,66,99,36,39,69 what zero's bring its as simple as that ! I will give examples below

Daytime 3 - 2 - 3
Evening 0 - 2 - 2

Daytime 5 - 7 - 6
Evening 0 - 4 - 0

Evening 1 - 6 - 9
Daytime 5 - 0 - 9

Evening 9 - 1 - 6
Daytime 8 - 5 - 9
Evening 2 - 4 - 0

****
82 and 28 the magic 8 ball studied

Evening 0 - 7 - 5
Daytime 4 - 7 - 8
Evening 4 - 6 - 8
Daytime 8 - 9 - 2
When 8x2 is wrote rest assured a zero will show up soon.

Daytime 0 - 0 - 5

Evening 9 - 3 - 4
Daytime 9 - 6 - 1
Evening 1 - 2 - 8
Daytime 5 - 2 - 4
When a 28 is written on the outside right hand your getting a double from one of the numbers below it see how the 500 showed up ? . Lets look at more examples .

Evening 9 - 1 - 1
Daytime 3 - 2 - 8
Evening 4 - 7 - 0
Sometimes it as easy as giving you addition 4+7 = 11 which is what showed up the next draw .

 Daytime 0 - 0 - 8
 Evening 2 - 1 - 8
and again another example of what I was talking about with 2x8 .

****

248 and 48 on the right hand side brings a double of one of the numbers below it

Evening 0 - 9 - 9
sometimes its quick

Daytime 2 - 4 - 8
Evening 5 - 9 - 5

Daytime 6 - 0 - 6
Evening 2 - 4 - 8
Daytime 5 - 6 - 8
and sometimes you have to wait for it

 Evening 1 - 0 - 0
 Daytime 5 - 9 - 3
 Evening 9 - 9 - 8

Daytime 2 - 4 - 8
Evening 6 - 4 - 1

****

This is another way to look at the pair system using mirrors to get the next pair faster than we normally would.

1 = 6

2 = 7

3 = 8

4 = 9

5 = 0

For example if the evening number was 439

Evening 4 - 3 - 9

4 = 9 we are going up one and down one so = 0

9

8

3 = 8 we are going up one and down one so = 9

8

7

9 =4 we are going up one and down one so 5

4

you likely got one or more pairs for this method than the old method of mirror pairs.Now take the pairs to the max ....from the first set of three we get/ sets 08,09, extreme side 00,99,88

from the second set we get 98,78,79,77,88,99 and from the last we get 54,53,43,55,44,33 maybe its not done you'd have to go back and look at 439 the last time it fell with some history to it . The number that fell behind that was 307 normalcy says that you waited for the pair

Under my box you didn't actually on my pair speed program you got the pairs by doing it a different way.

****

413 will produce the 12 set , also it introduces the 11 pair which as we all know they do not like to give out . Most of the time 413 will give you 162 in some form before its all done it will also give you doubles 11,44,33, but when you hit the 11 pair or 162 consider the ride over......can you hit multiple doubles from this number the answer is yes

Can you get another 13 set that does that same thing yes you can I consider all the 13 sets the same 013,213,413,513,613,713,813,913 notice I did not put the doubles of it in there yes all of them do the same thing you get a 12 out of them and when you hit a double from that 112, or the number like I just said comes back with a double from the number take 513 example 55, 11, 33, Remember this is for 13 and has to fall just like that

****
When one of these numbers fall another will fall from the group.

227

222

332

332

272

323

733

373

****

The power of 10

you'll need to know what equates to 10 so here are the numbers

00 = 10 / 05

55= 10

91= 10

82= 10

73= 10

64= 10

the double that would equate to ten

442 622

So every time you see a ten you will see another ten in some form or fashion usually this goes on for 3 draw string

Lets be aware that 10 usually starts on the inside and continues on going to try and find some examples

Evening	7 - 8 - 2
Daytime	0 - 1 - 9
Evening	8 - 1 - 0

810 Lets look at it again

Daytime	7 - 3 - 2
Evening	7 - 0 - 1
Daytime	7 - 0 - 5

We started with 705 we know that 0=5 and gave back 701 and then 732 so we had 3 10's in a row .

Going to a different month

Evening	5 - 5 - 8
Daytime	5 - 5 - 1
Evening	
	6 - 5 - 0

Again with the 05 pair , 55 pair and another 55 which all equates to the number 10

Evening	6 - 2 - 2
Daytime	9 - 1 - 6

Another look at how ten is brought back .

I hope this will help turn a light on for some and realize that ole way pair systems are okay but it takes many systems to get a number

****

Have you ever noticed the magic of number 1 We are going to look at number 1 on the outside written like this x-x-1 the number 1 can set off a chain of events that are something . I'm going to give multiple examples here to show what I'm talking about. Btw this is not done with a series numbers such as 901, 201 why b/c those numbers usually set off different events .

5 - 9 - 9

8 - 0 - 1

the number is written as with the 1 on the outside ....

Daytime   1 - 0 -5

Evening   4 - 0 - 1

the one here above gives you 4+1=5 and joins together the 1 and 0

Daytime   7 - 6 - 9

Evening   4 - 6 - 1

we got 461 is 16 next draw we got 7+9 =16

Daytime   6 - 1 - 4

Evening   3 - 6 - 1

Simply giving us the 16 back next draw and joining the 3 and making it 3+1=4

Was trying to find a good month for this so we're going to skip ahead

Daytime   9 - 9 - 2

Evening   7 - 3 - 1

Daytime   6 - 3 - 1

6+3+1 = 10 which equates to 7+3= same throwing another one on the outside they added the 1 and gave you a double with 992 which is 9+2=11 same thing as the number below it 7+3+1 = 11 same as your double

Daytime   4 - 6 - 2

Evening   9 - 7 - 1

7+1 = 8 next draw 6+2 =8 ....joining the 9 which evens it out to 10 written below and above . 9+1=10 and 4+6=10 ....now that I have fully made you think I hope your looking forward to the next 1 on the outside.

****
What's special about 439 not a lot but you will need to know your pairs to get through the next few draws.

We're going to actually look at the number 12 and 13 pairs

12 pairs are 12 (simply gives you the whole number)

552 a double that gives you 12

93x

84x

75x

66x

002 is another way you can write twelve 0=5

or sum 12

****

13 pairs are 13 (again simply gives you the whole number)

677 or 667 using the double to get you to 13

94x

85x

76x (just as the same double example I gave above)

003 (just using the 0=5 method)

553

or sum 13
****

479 047 125 146 167 235 245 248 249 257 278 279 467 478
012 015 018 019 023 024 027 035 038 039 044 045 048 049 057 077
078 079 123 128 129 135 138 139 158 159 224 227 228 229 234 237
238 239 247 289 345 348 349 357 358 359 378 379 446 448 449 458
459 489 578 579 677 778 779 789

These numbers follow each other when one falls another will come within the next few draws.

****

087 in any form sets off a double sum 15 you should look for 663,055,005,100,933,177,447,555

****

707 circle 02,07,52,57 pairs
002,007,052,057,502,507,552,557
012,017,062,067,512,517,562,567
022,027,072,077,522,527,572,577
023,028,073,078,523,528,573,578
024,029,074,079,524,529,574,579

****

These are doubles that like to fall behind each other see one and you will see another soon.
330, 331, 332, 333, 334, 335, 336, 337, 338, 339,
660, 661, 662, 663, 664, 665, 666, 667, 668, 669,
990, 991, 992, 993, 994, 995, 996, 997, 998, 999
556, 559, 166, 566, 669, 199, 599, 699
****

I call these leader numbers not always but most of the time you will get a pair from the leader number . You find the pair by looking at all leader numbers and picking your pair out of the group. This is more of a guide lesson

0 =00, 19,28,37,46 ,00, 55, 64, 73

1 =01,10,29,38,47, 00, 56, 66, 75

2 =02,11,20,39,48,00, 58, 67,46,26,78,48,47

3 =03,12,21,30,49,00, 59, 68,46,26 78,48,47

4 =04,13,22,31,40,50, 00, 69,67,46

5 =05,14,23,32,41;60

6 =06,15,24,33,42,61

7 =07,16,25,34,43,52, 61,70,67,46

8 =08,17,26,35,44,53,62,71,67,46

9 =09,18,27,36,45,54, 63, 72,46,26 78,48,47

****
The number 3 in pick 3

         Evening   8 - 1 - 0
         Daytime   9 - 5 - 3
See how that 3 + 5 came back as an 8 last month

         Evening   7 - 5 - 2
         Daytime   4 - 1 - 3

Actually this time they added 4+1 = 5 / 1-3=2 / 4+3 =7 result 752

         Daytime   6 - 0 - 5
         Evening
6 - 6 - 3

best way to describe this they are going 6-6 =0 / mirror to 0 = 5 / 6-3=3 +6-3=3 for the whole 6 = 605

Evening   2 - 8 - 8
Daytime   3 - 6 - 3
yes I'm going to attempt to explain this one lol ....here we
go

6 dived by 2 = 3 / 3 x 6 = 18 which is 8+8 = 16 + 2 = 18
****
When one falls another will fall very quickly watch them
closely.

029 038 056 057 058 068 078 089

135 139 145 149 156 157 159 169

235 236 237 239 245 246 247 249 256 257 259 269 279

348 356 357 358 368 369 378

457 458 468

****
We're going to call this the ones that can't leave each other alone
its a rare bunch but sometimes when they show up they show
up with their own kind.....oh sure you got the pairs like I just
showed you but these come along and they seems to gravitate
towards each other when they really shouldn't so they are
together more times than not.

37x = 1,3,5,7,9

19x = 1,3,5,7,9

46x= 2,4,6,8,0

28x =2,4,6,8,0

39x= 1,3,5,7,9

48x= 2,4,6,8,0

57x=1,3,5,7,9

00x = 2,4,6,8,0

11x =1,3,5,7,9

22x =2,4,6,8,0

77x=1,3,5,7,9

88x =2,4,6,8,0

99x=1,3,5,7,9

33x=1,3,5,7,9

44x =2,4,6,8,0

55x=1,3,5,7,9

66x=2,4,6,8,0

****
Below you find the pairs that usually go with each other and the x is the opposite . So remember usually two positives are completed with a negative and two negatives are completed with a positive position number..........so are you keeping up with your ex by the way its not cost effective to play all them I just wanted to give you another way to look at numbers .
02 x = 1,3,5,7,9
04 x = 1,3,5,7,9
13 x = 0,2,4,6,8
59 x = 0,2,4,6,8
68 x = 1,3,5,7,9

79 x = 0,2,4,6,8
06 x = 1,3,5,7,9
24 x = 1,3,5,7,9
15 x = 0,2,4,6,8
08 x = 1,3,5,7,9
17 x = 0,2,4,6,8
35 x = 0,2,4,6,8

****

When 704 hits straight like this then you will get a follow up 00, 77, 44, probably not in that order but at least two doubles will come from it. The exception to this rule would be if it hit late in the month .

70x will produce these pairs usually within a few draws does not matter if its front or back 70x pair

24x 70x 27x 14x 45x 48x 22x 20x 08x 18x 16x 78x 23x

The exception to the rule would be if you get a pair of these prior (the draw before) then this does not usually work.
****

09,19,29,39,49,59,69,79,89,99

PAIR USUALLY BRINGS

05,15,25,35,45,65,75,85,95

830/ when 830 shows a double of 00/11/22/33/44/55/66/77/88/99 will show soon look at the immediate double that fell the last time this number fell and you will go up one or down to get the next double so if the number was a 499 then you will look for

a 55/00 double to show next time....research you will
see what i mean

****

**Lets look at the number that fell last night 838 can we use the
doubler system yes it will stop when you get a 76/67 pair either in
a double or a single version.**

**This number sets off pairs that will fall and bounce off each other
all week . The pairs for the entire week and there may be a draw it
skips so play play play what you feel is coming....**

**The 9 pair set that is two numbers that equate to 9 =9,
09,81,72,63,54, outside the box 04x (we treat 0 as 5)**

**The 10 pair set = 10,55,91,00,64,82,73 and the most important
set of all 11 pair set b/c you got an 11 last night with 83 so**

**011, 55, 92, 11, 83,74,65 (do not stop this set just b/c they throw
you a sum 11 it will continue on) lets look**

Daytime	**4 - 7** - 8
Evening	3 - **9 - 2**
Daytime	4 - **1 - 9**
Evening	1 - 4 - 3
Evening	**9** - 8 - 8
Daytime	2 - **4 - 0**
Evening	**9 - 9** - 7

Daytime	4 - 8 - **9**
Evening	**4** - 1 - **7**
Daytime	4 - **9 - 9**
Evening	**8 - 3 - 0**
Daytime	2 - **0 - 6**
Evening	**0 - 1** - 6
Daytime	0 **- 2 - 9**
Evening	8 - 3 - 8

One of these will drop and then a single will drop or vice verse

001/011,
119/199,
227/277,
335/355,
344/334,
155/115,
669/699,
588/558,
399/339

singles
see
you
watch
these

056,
057,
058,
059,
067,
068,
069,
078,
079,
089,

567,
568,
569,
578,
579,
589,
678,
679,
689, 789